MASTER

FRACTIONS!

SECOND EDITION

Abdulaziz. M. Alibarre, Ph.D.

Contents

WELCOME STUDENTS AND TEACHERS!

Fractions are very important in math, science and in everyday life. For example great chefs should have a good grasp of fractions to cook great meals. They should know the exact proportions or fractions of ingredients they need to mix. Also, pharmacists need to know fractions well so as to measure the correct dosage of medicine. Mastering the fractions will help you to excel in math including algebra and beyond.

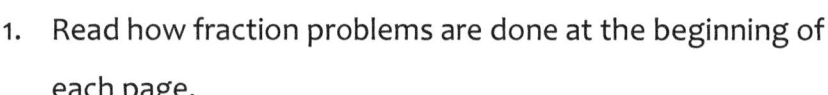

Just follow my advice and you will complete your fraction in no time:

1. Read how fraction problems are done at the beginning of each page.
2. See and understand the examples.
3. Work on problems.
4. Check your answers at the end of the book!
5. Do the problem again if you do not get the answer first time.

That is it folks! See you in Pre-algebra!

1) HOW TO NAME FRACTIONS

Fraction is a part of a whole object.
- ✓ When 5 people share a pizza equally each will get one fifth
- ✓ The top number is called **Numerator** and the lower number is called **Denominator**.

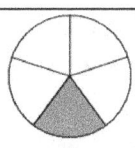

$$\frac{1}{5}$$

1) What fractions is shaded?

2) What fractions is shaded?

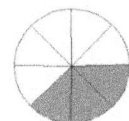

3) What fraction is not shaded?.

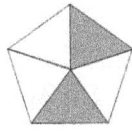

4) What fraction is letter (v) in:

   Love

5) What fraction is letter (**n**) in:

   Minneapolis

6) Shade 7/10

7) Shade 5/10.

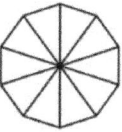

8) Shade 9/10.

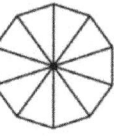

9) Shade 4/4.

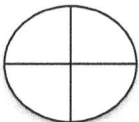

10) What fraction is letter (u) in:

   Suuban

2) IDENTIFY PROPER, IMPROPER, AND MIXED FRACTIONS

A proper fraction is less than 1 or has the numerator smaller than denominator	$\dfrac{3}{7}$
An improper fraction is equal or more than 1 (usually, the numerator is bigger than denominator).	$\dfrac{7}{3},\quad \dfrac{5}{5}=1$
A mixed fraction is a whole number and proper faction combined:	$6\dfrac{2}{3}$

Write Proper **(P)**, Improper **(I)**, of Mixed Fractions **(MF)**

1) $\dfrac{3}{3}$

2) $\dfrac{9}{4}$

3) $\dfrac{8}{3}$

4) $5\dfrac{1}{4}$

5) $\dfrac{7}{4}$

6) $6\dfrac{1}{2}$

7) $1\dfrac{1}{2}$

8) $\dfrac{5}{9}$

9) $\dfrac{7}{9}$

10) $\dfrac{12}{12}$

11) $6\dfrac{1}{2}$

12) $\dfrac{5}{6}$

3) EQUIVALENT FRACTIONS: WRITE THE MISSING NUMBER

Example: Write the missing number to get equivalent fractions: $\dfrac{2}{4} = \dfrac{6}{?}$

Step 1) *See that the 2 times 3 is what gives us the 6.* $\dfrac{2^{\times 3}}{4} = \dfrac{6}{?}$

Step2) *So, multiply the 3 also to the 4 to get 12* $\dfrac{2^{\times 3}}{4_{\times 3}} = \dfrac{6}{12}$

---

1) $\dfrac{2}{4} = \dfrac{8}{?}$

2) $\dfrac{2}{1} = \dfrac{?}{5}$

3) $\dfrac{2}{3} = \dfrac{8}{?}$

4) $\dfrac{2}{5} = \dfrac{?}{15}$

5) $\dfrac{4}{10} = \dfrac{?}{20}$

6) $\dfrac{4}{9} = \dfrac{?}{45}$

7) $\dfrac{8}{20} = \dfrac{?}{5}$

8) $\dfrac{21}{33} = \dfrac{?}{11}$

9) $\dfrac{15}{18} = \dfrac{?}{6}$

10) $\dfrac{1}{3} = \dfrac{8}{?}$

11) $\dfrac{5}{2} = \dfrac{25}{?}$

12) $\dfrac{14}{49} = \dfrac{2}{?}$

13) $\dfrac{7}{8} = \dfrac{28}{?}$

14) $\dfrac{4}{7} = \dfrac{100}{?}$

15) $\dfrac{14}{41} = \dfrac{56}{?}$

Example : Write the missing number to get equivalent fractions: $\dfrac{4}{?} = \dfrac{16}{20}$

Step 1: Work backwards: 16 divide 4 is 4?	$\dfrac{4}{?} = \dfrac{16 \div 4}{20}$
Step 2: Divide the 20 by 4 also to get 5.	$\dfrac{4}{5} = \dfrac{16 \div 4}{20 \div 4}$

1) $\dfrac{2}{?} = \dfrac{8}{20}$

2) $\dfrac{7}{?} = \dfrac{14}{10}$

3) $\dfrac{2}{?} = \dfrac{16}{24}$

4) $\dfrac{?}{5} = \dfrac{100}{25}$

5) $\dfrac{?}{5} = \dfrac{38}{10}$

6) $\dfrac{1}{?} = \dfrac{7}{28}$

7) $\dfrac{2}{?} = \dfrac{18}{72}$

8) $\dfrac{21}{?} = \dfrac{42}{22}$

9) $\dfrac{5}{?} = \dfrac{30}{36}$

10) $\dfrac{1}{?} = \dfrac{8}{64}$

11) $\dfrac{5}{?} = \dfrac{25}{10}$

12) $\dfrac{?}{7} = \dfrac{70}{49}$

13) $\dfrac{7}{?} = \dfrac{28}{8}$

14) $\dfrac{4}{?} = \dfrac{100}{25}$

15) $\dfrac{8}{?} = \dfrac{56}{42}$

5) COMPARE TWO FRACTIONS

First, remember the symbols:
> means greater than ;   Eight is greater than 5 is written as 8 > 5
< means less;   Five is less than nine is written as 5 < 9

Example : Compare Fractions $\qquad$ $\dfrac{2}{5} \bigcirc \dfrac{3}{4}$

Step 1: Cross Multiply *the opposite numbers:*	$2 \times 4 < 5 \times 3$	$\Rightarrow$	$\dfrac{2}{5} < \dfrac{3}{4}$
Step 2: Decide: *Since **8** is less and is on the left side, the fraction on the left side is smaller*	$8 < 15$		

1) $\dfrac{2}{5} \bigcirc \dfrac{3}{7}$ 　　　 2) $\dfrac{7}{9} \bigcirc \dfrac{5}{7}$ 　　　 3) $\dfrac{1}{2} \bigcirc \dfrac{5}{7}$

4) $\dfrac{4}{9} \bigcirc \dfrac{7}{13}$ 　　　 5) $\dfrac{11}{12} \bigcirc \dfrac{8}{9}$ 　　　 6) $\dfrac{2}{7} \bigcirc \dfrac{3}{8}$

7) $\dfrac{2}{7} \bigcirc \dfrac{4}{14}$ 　　　 8) $\dfrac{7}{9} \bigcirc \dfrac{5}{7}$ 　　　 9) $\dfrac{3}{5} \bigcirc \dfrac{5}{6}$

10) $\dfrac{2}{15} \bigcirc \dfrac{3}{13}$ 　　　 11 $\dfrac{2}{18} \bigcirc \dfrac{3}{27}$ 　　　 12) $\dfrac{5}{9} \bigcirc \dfrac{5}{8}$

13) $\dfrac{1}{5} \bigcirc \dfrac{3}{4}$ 　　　 14 $\dfrac{11}{12} \bigcirc \dfrac{23}{24}$ 　　　 15) $\dfrac{2}{3} \bigcirc \dfrac{3}{4}$

6) SIMPLIFY PROPER FRACTIONS: METHOD 1

Example: Simplify completely: $\dfrac{16}{24}$

Divide both the denominator and numerator by the *same numbers until you can't divide it anymore.*	$\dfrac{16^{\div 4}}{24_{\div 4}} = \dfrac{4^{\div 2}}{6_{\div 2}} = \dfrac{2}{3}$
A quicker way is to divide it to the common factor:	$\dfrac{16^{\div 8}}{24_{\div 8}} = \dfrac{2}{3}$

1) $\dfrac{2}{4} =$ 2) $\dfrac{2}{16} =$ 3) $\dfrac{12}{14} =$

4) $\dfrac{10}{25} =$ 5) $\dfrac{24}{60} =$ 6) $\dfrac{6}{27} =$

7) $\dfrac{30}{35} =$ 8) $\dfrac{21}{33} =$ 9) $\dfrac{9}{36} =$

10) $\dfrac{12}{30} =$ 11) $\dfrac{35}{42} =$ 12) $\dfrac{3}{27} =$

13) $\dfrac{35}{49} =$ 14) $\dfrac{3}{54} =$ 15) $\dfrac{3}{21} =$

16) $\dfrac{10}{30} =$ 17) $\dfrac{21}{75} =$ 18) $\dfrac{15}{45} =$

7) SIMPLIFY PROPER FRACTIONS: METHOD 2

Examples:
$$\frac{16}{24} = \frac{\cancel{8} \times 2}{\cancel{8} \times 3} = \frac{2}{3} \qquad\qquad \frac{35}{49} = \frac{\cancel{7} \times 5}{\cancel{7} \times 7} = \frac{5}{7}$$

1) $\dfrac{4}{18} =$

2) $\dfrac{5}{35} =$

3) $\dfrac{3}{15} =$

4) $\dfrac{12}{42} =$

5) $\dfrac{24}{50} =$

6) $\dfrac{12}{27} =$

7) $\dfrac{8}{22} =$

8) $\dfrac{14}{49} =$

9) $\dfrac{9}{18} =$

10) $\dfrac{14}{30} =$

11) $\dfrac{35}{42} =$

12) $\dfrac{6}{27} =$

13) $\dfrac{25}{30} =$

14) $\dfrac{35}{100} =$

15) $\dfrac{30}{42} =$

8) SIMPLIFY MIXED NUMBERS

Examples:

$$4\frac{4}{12} = 4\frac{4 \div 4}{12 \div 4} = 4\frac{1}{3} \qquad\qquad 6\frac{14}{21} = 6\frac{14 \div 7}{21 \div 7} = 6\frac{2}{3}$$

1) $3\frac{3}{12} =$

2) $12\frac{5}{15} =$

3) $6\frac{4}{10} =$

4) $7\frac{6}{18} =$

5) $8\frac{8}{10} =$

6) $9\frac{4}{16} =$

7) $9\frac{3}{9} =$

8) $5\frac{7}{14} =$

9) $21\frac{14}{28} =$

10) $8\frac{8}{18} =$

11) $12\frac{4}{12} =$

12) $14\frac{8}{10} =$

13) $9\frac{4}{10} =$

14) $15\frac{15}{25} =$

15) $11\frac{10}{12} =$

9) CHANGE IMPROPER FRACTIONS INTO MIXED NUMBERS

To change improper fractions into mixed numbers, divide and write your answer as shown.	$\dfrac{10}{3}$ $\Rightarrow$ $3\overline{)10}$ gives 3 remainder, $\dfrac{9}{1}$ $\Rightarrow$ $3\dfrac{1}{3}$

1) $\dfrac{5}{4}$

2) $\dfrac{13}{9}$

3) $\dfrac{27}{8}$

4) $\dfrac{61}{6}$

5) $\dfrac{15}{2}$

6) $\dfrac{25}{3}$

7) $\dfrac{19}{3}$

8) $\dfrac{51}{4}$

9) $\dfrac{71}{8}$

10) $\dfrac{42}{8}$

11) $\dfrac{73}{32}$

12) $\dfrac{87}{13}$

13) $\dfrac{25}{17}$

14) $\dfrac{162}{50}$

15) $\dfrac{97}{31}$

10) SIMPLIFY IMPROPER FRACTIONS

First simplify, then change the improper fractions into mixed numbers	$\dfrac{18}{10} = \dfrac{9}{5}$ $\Rightarrow$ $5\overline{)9}$ $\underline{5}$ 4 $\Rightarrow$ $1\dfrac{4}{5}$

1) $\dfrac{15}{6}$

2) $\dfrac{16}{10}$

3) $\dfrac{18}{12}$

4) $\dfrac{26}{4}$

5) $\dfrac{80}{30}$

6) $\dfrac{100}{40}$

7) $\dfrac{85}{15}$

8) $\dfrac{52}{24}$

9) $\dfrac{72}{8}$

10) $\dfrac{75}{27}$

11) $\dfrac{88}{55}$

12) $\dfrac{56}{16}$

13) $\dfrac{25}{10}$

14) $\dfrac{62}{50}$

15) $\dfrac{81}{36}$

11) SIMPLIFY IMPROPER MIXED NUMBERS

| First: Simplify the improper if possible Next: change the improper fractions into mixed numbers Finally: add the mixed and the whole number | $5\dfrac{10}{6} = 5\dfrac{5}{3}$ | $3\overline{)5}$ $\dfrac{3}{2}$ $\begin{array}{r}1\\\end{array}$ $\Longrightarrow$ | $5 + 1\dfrac{2}{3} = 6\dfrac{2}{3}$ |

1) $5\dfrac{10}{3}$

2) $6\dfrac{17}{10}$

3) $14\dfrac{18}{2}$

4) $8\dfrac{16}{4}$

5) $9\dfrac{8}{3}$

6) $10\dfrac{10}{7}$

7) $6\dfrac{15}{7}$

8) $8\dfrac{42}{4}$

9) $13\dfrac{50}{40}$

10) $2\dfrac{25}{10}$

11) $1\dfrac{35}{15}$

12) $21\dfrac{56}{16}$

13) $4\dfrac{45}{9}$

14) $5\dfrac{60}{50}$

15) $16\dfrac{90}{81}$

Example: Change $5\frac{1}{4}$ into improper fraction

Step 1: Multiply the denominator with the whole number	$5 \times 4 = 20$
Step 2: Add the product to the numerator.	$20 + 1 = 21$
Step 3: Keep your original denominator and write the results	$\frac{21}{4}$

1) $3\frac{1}{2} =$

2) $5\frac{2}{3} =$

3) $8\frac{1}{4} =$

4) $8\frac{1}{3} =$

5) $9\frac{2}{7} =$

6) $8\frac{6}{7} =$

7) $4\frac{3}{7} =$

8) $4\frac{5}{9} =$

9) $3\frac{5}{14} =$

10) $3\frac{5}{9} =$

11) $4\frac{3}{5} =$

12) $4\frac{5}{12} =$

13) $4\frac{3}{8} =$

14) $5\frac{5}{7} =$

15) $8\frac{7}{12} =$

A) Compare Fractions: Fill (<, >, or =):

1) $\dfrac{3}{5}$ $\dfrac{5}{7}$ 2) $\dfrac{1}{3}$ $\dfrac{3}{5}$ 3) $\dfrac{2}{3}$ $\dfrac{2}{7}$ 4) $\dfrac{1}{3}$ $\dfrac{3}{4}$

B) Simplify:

5) $\dfrac{14}{80} =$ 6) $\dfrac{10}{68} =$ 7) $\dfrac{44}{55} =$ 8) $\dfrac{36}{90} =$

C) Change into Improper Fractions:

9) $3\dfrac{1}{2} =$ 10) $5\dfrac{2}{3} =$ 11) $7\dfrac{2}{9} =$ 12) $8\dfrac{1}{2} =$

D) Change into Mixed Numbers:

13) $\dfrac{5}{4} =$ 14) $\dfrac{13}{9} =$ 15) $\dfrac{27}{8} =$ 16) $\dfrac{19}{7} =$

E) Change into Mixed Numbers:

17) $5\dfrac{5}{3} =$ 18) $6\dfrac{9}{5} =$ 19) $8\dfrac{20}{6} =$ 20) $5\dfrac{12}{7} =$

14) ADD PROPER FRACTION AND A WHOLE NUMBER

	Example 1	Example 2
	$\frac{1}{2} + 4 \Downarrow$	$5 + \frac{3}{4} \Downarrow$
Just combine the fraction with the whole number. That is it!	$4\frac{1}{2}$	$5\frac{3}{4}$

1) $\frac{1}{4} + 2$

2) $\frac{2}{3} + 5$

3) $\frac{3}{10} + 3$

4) $7 + \frac{3}{8}$

5) $3 + \frac{2}{3}$

6) $\frac{7}{9} + 5$

7) $\frac{1}{5} + 6$

8) $\frac{8}{15} + 5$

9) $10 + \frac{3}{11}$

10) $4 + \frac{3}{4}$

11) $9 + \frac{3}{11}$

12) $\frac{5}{28} + 8$

13) $6 + \frac{1}{2}$

14) $9 + \frac{5}{19}$

15) $8 + \frac{2}{13}$

15) ADD IMPROPER FRACTION AND A WHOLE NUMBER

	Example 1	**Example 2**
Simplify these Fractions:	$\dfrac{9}{2} + 3$	$5 + \dfrac{12}{7}$

Step 1. Change any improper fractions into mixed numbers (Simplify if needed)

Step 2. Add the whole numbers.

Example 1: $\dfrac{9}{2} = 4\dfrac{1}{2}$ $\qquad 4\dfrac{1}{2} + 3 = 7\dfrac{1}{2}$

Example 2: $\dfrac{12}{7} = 1\dfrac{5}{7}$ $\qquad 5 + 1\dfrac{5}{7} = 6\dfrac{5}{7}$

1) $\dfrac{4}{3} + 2$

2) $\dfrac{3}{2} + 5$

3) $\dfrac{10}{3} + 3$

4) $7 + \dfrac{8}{3}$

5) $3 + \dfrac{5}{3}$

6) $\dfrac{5}{2} + 5$

7) $\dfrac{12}{5} + 6$

8) $\dfrac{15}{8} + 5$

9) $10 + \dfrac{11}{3}$

10) $4 + \dfrac{15}{4}$

11) $9 + \dfrac{7}{6}$

12) $\dfrac{28}{5} + 8$

13) $6 + \dfrac{17}{5}$

14) $9 + \dfrac{19}{5}$

15) $8 + \dfrac{13}{2}$

16) ADD PROPER FRACTIONS WITH LIKE DENOMINATORS

Example: Simplify the fractions $\frac{3}{10} + \frac{9}{10}$

1. Add the numerators $3 + 9 = 12$ $\Longrightarrow$	2. Keep the denominators unchanged. $= \frac{12}{10}$ $\Longrightarrow$	3) Simplify/rename as needed! $\frac{12}{10} = \frac{6}{5} = 1\frac{1}{5}$

Very important: Never add or subtract the denominators!

1) $\frac{1}{5} + \frac{2}{5}$

2) $\frac{5}{9} + \frac{4}{9}$

3) $\frac{2}{7} + \frac{3}{7}$

4) $\frac{5}{9} + \frac{5}{9}$

5) $\frac{3}{4} + \frac{1}{4}$

6) $\frac{7}{8} + \frac{3}{8}$

7) $\frac{3}{4} + \frac{3}{4}$

8) $\frac{3}{8} + \frac{1}{8}$

9) $\frac{5}{7} + \frac{3}{7}$

10) $\frac{1}{4} + \frac{1}{4}$

11) $\frac{7}{10} + \frac{8}{10}$

12) $\frac{9}{21} + \frac{5}{21}$

13) $\frac{5}{12} + \frac{1}{12}$

14) $\frac{7}{25} + \frac{3}{25}$

15) $\frac{6}{11} + \frac{7}{11}$

Example: Simplify the fractions $\frac{7}{5} + \frac{9}{5}$

1. Add the numerators (top numbers)	2. Keep the denominator unchanged.	3) Simplify/rename the fraction as needed!
$7 + 9 = 16$ $\Longrightarrow$	$= \dfrac{16}{5}$ $\Longrightarrow$	$= \dfrac{16}{5} = 3\dfrac{1}{5}$

1) $\dfrac{6}{5} + \dfrac{7}{5}$

2) $\dfrac{10}{7} + \dfrac{9}{7}$

3) $\dfrac{12}{5} + \dfrac{9}{5}$

4) $\dfrac{8}{3} + \dfrac{7}{3}$

5) $\dfrac{7}{4} + \dfrac{9}{4}$

6) $\dfrac{8}{7} + \dfrac{10}{7}$

7) $\dfrac{11}{6} + \dfrac{7}{6}$

8) $\dfrac{9}{8} + \dfrac{11}{8}$

9) $\dfrac{15}{7} + \dfrac{11}{7}$

10) $\dfrac{20}{9} + \dfrac{11}{9}$

11) $\dfrac{9}{5} + \dfrac{9}{5}$

12) $\dfrac{14}{13} + \dfrac{15}{13}$

13) $\dfrac{13}{10} + \dfrac{11}{10}$

14) $\dfrac{17}{12} + \dfrac{13}{12}$

15) $\dfrac{10}{11} + \dfrac{15}{11}$

18) SUBTRACT PROPER FRACTIONS: LIKE DENOMINATORS

Example: Simplify the fractions $\frac{3}{10} - \frac{1}{10}$

1. **Subtract** the numerators (top numbers) $3 - 1 = 2$	2. **Keep the denominator** unchanged. $\frac{3-1}{10} = \frac{2}{10}$	3) **Simplify/Rename** the fraction as needed. $\Rightarrow \quad = \frac{1}{5}$

Very important Reminder: *Never add or subtract the denominators. Also, make sure to simplify fractions every time!*

1) $\frac{2}{5} - \frac{1}{5}$	6) $\frac{7}{8} - \frac{3}{8}$	11) $\frac{5}{12} - \frac{3}{12}$
2) $\frac{6}{5} - \frac{4}{5}$	7) $\frac{3}{4} - \frac{3}{4}$	12) $\frac{9}{11} - \frac{3}{11}$
3) $\frac{6}{7} - \frac{3}{7}$	8) $\frac{3}{8} - \frac{1}{8}$	13) $\frac{3}{10} - \frac{1}{10}$
4) $\frac{5}{9} - \frac{5}{9}$	9) $\frac{7}{7} - \frac{3}{7}$	14) $\frac{9}{14} - \frac{5}{14}$
5) $\frac{7}{9} - \frac{5}{9}$	10) $\frac{8}{9} - \frac{7}{9}$	15) $\frac{17}{18} - \frac{7}{18}$

Simplify. Make sure to reduce and to rename the fractions as needed.

1) $\dfrac{1}{3} + 5$

2) $9 + \dfrac{4}{3}$

3) $\dfrac{11}{5} + \dfrac{12}{5}$

4) $7 + \dfrac{14}{4}$

5) $6 + \dfrac{12}{5}$

6) $\dfrac{3}{2} + \dfrac{7}{2}$

7) $\dfrac{12}{5} + \dfrac{6}{5}$

8) $\dfrac{1}{8} + \dfrac{5}{8}$

9) $\dfrac{9}{14} - \dfrac{3}{14}$

10) $\dfrac{12}{5} + \dfrac{19}{5}$

11) $\dfrac{11}{3} + 5$

12) $\dfrac{5}{2} + \dfrac{12}{5}$

13) $10 + \dfrac{11}{3}$

14) $\dfrac{8}{15} - \dfrac{5}{15}$

15) $11 + \dfrac{11}{3}$

Example: Simplify the fractions $5\frac{3}{10} + 3\frac{1}{10}$

1. Add the whole Numbers.	2. Add the numerators and *keep the denominator* unchanged.	3) Combine the whole number and the simplified fractions.
$5 + 3 = 8$	$\frac{3+1}{10} = \frac{4}{10} = \frac{2}{5}$	$8 + \frac{2}{5} = 8\frac{2}{5}$

1) $2\frac{3}{5} + 3\frac{1}{5}$

2) $5\frac{2}{7} + 2\frac{3}{7}$

3) $4\frac{2}{8} + 1\frac{1}{8}$

4) $6\frac{3}{7} + 3\frac{1}{7}$

5) $1\frac{6}{11} + 2\frac{3}{11}$

6) $1\frac{1}{5} + 3\frac{2}{5}$

7) $7\frac{5}{12} + 12\frac{1}{12}$

8) $7\frac{5}{16} + 2\frac{7}{16}$

9) $1\frac{1}{5} + 3\frac{2}{5}$

10) $4\frac{5}{18} + 3\frac{1}{18}$

11) $5\frac{1}{4} + 2\frac{1}{4}$

12) $3\frac{2}{7} + 2\frac{1}{7}$

13) $5\frac{11}{15} + 1\frac{3}{15}$

14) $6\frac{2}{5} + 5\frac{1}{5}$

15) $1\frac{14}{95} + 2\frac{6}{95}$

Example: Simplify the fractions $5\frac{4}{5} + 3\frac{3}{5}$

2. Add the whole Numbers.	2. Add the numerators/ *keep the denominator* unchanged.	3) Combine the whole number and the mixed numbers.
$5 + 3 = 8$	$\frac{7}{5} = 1\frac{2}{5}$	$8 + 1\frac{2}{5} = 9\frac{2}{5}$

 Look how we have renamed the $\frac{7}{5}$ into mixed numbers $1\frac{2}{5}$ and then added to the 8

1) $1\frac{3}{5} + 3\frac{2}{5}$	2) $1\frac{5}{6} + 3\frac{1}{6}$	3) $8\frac{5}{9} + 9\frac{4}{9}$	4) $6\frac{3}{4} + 7\frac{1}{4}$
5) $5\frac{4}{7} + 2\frac{5}{7}$	6) $7\frac{5}{12} + 2\frac{11}{12}$	7) $8\frac{4}{5} + 2\frac{3}{5}$	8) $3\frac{5}{7} + 2\frac{4}{7}$
9) $4\frac{7}{8} + 1\frac{1}{8}$	10) $7\frac{5}{6} + 2\frac{5}{6}$	11) $5\frac{7}{13} + 1\frac{8}{13}$	12) $5\frac{11}{15} + 1\frac{4}{15}$

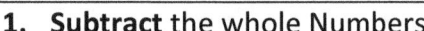

1. **Subtract** the whole Numbers. 2. **Subtract** the numerators 3. **Keep** the denominator unchanged. 4. **Simplify/rename** as needed.	$Example 1)\ 9\frac{5}{6} - 5\frac{1}{6} = \quad 4\frac{4}{6} = 4\frac{2}{3}$
	$Example\ 2)\ 9\frac{7}{5} - 3\frac{2}{5} \quad 6\frac{5}{5} = 6 + 1 = 7$

 Look how we changed the $\frac{5}{5}$ **into a 1** *and added it to the 6*

1) $7\frac{3}{5} - 3\frac{2}{5}$	2) $6\frac{5}{6} - 3\frac{1}{6}$	3) $8\frac{5}{9} - 5\frac{4}{9}$	4) $6\frac{3}{4} - 2\frac{1}{4}$
5) $5\frac{6}{7} - 2\frac{5}{7}$	6) $12\frac{5}{12} - 5\frac{1}{12}$	7) $8\frac{4}{5} - 2\frac{3}{5}$	8) $3\frac{5}{7} - 2\frac{4}{7}$
9) $4\frac{7}{8} - 1\frac{1}{8}$	10) $7\frac{5}{6} - 2\frac{5}{6}$	11) $5\frac{17}{11} - 1\frac{6}{11}$	12) $5\frac{11}{15} - 1\frac{4}{15}$

23) SUBTRACT MIXED NUMBERS: IMPROPER FRACTION METHOD

▪ See we couldn't subtract the numerators $(3-5)$ ▪ So, change the mixed numbers into improper fraction ▪ Subtract and Simplify.	**Simplify:** $5\dfrac{3}{6} - 1\dfrac{5}{6}$ $\dfrac{33}{6} - \dfrac{11}{6} = \dfrac{22}{6} = \dfrac{11}{3} = 3\dfrac{2}{3}$

1) $5\dfrac{3}{5} - 3\dfrac{4}{5}$

2) $7\dfrac{1}{5} - 3\dfrac{2}{5}$

3) $7\dfrac{2}{5} - 1\dfrac{3}{5}$

4) $5\dfrac{1}{7} - 2\dfrac{3}{7}$

5) $5\dfrac{1}{13} - 1\dfrac{3}{13}$

6) $4\dfrac{4}{15} - 1\dfrac{7}{15}$

7) $14\dfrac{2}{8} - 11\dfrac{7}{8}$

8) $7\dfrac{4}{16} - 2\dfrac{7}{16}$

9) $5\dfrac{3}{10} - 1\dfrac{7}{10}$

10) $15\dfrac{3}{7} - 3\dfrac{4}{7}$

11) $7\dfrac{1}{5} - 3\dfrac{2}{5}$

12) $9\dfrac{2}{9} - 5\dfrac{3}{9}$

Renaming whole numbers is useful in the addition and specially the subtraction of the fractions. Let's practice now:

To rename a whole number subtract 1 form the whole number and then change the 1 into any fractions you like. ***Make sure the numerator and the denominator are equal***: See we could rename the 7 in many ways depending our need. Could you think 4 more ways to rename the 7?

$$7 = 6\frac{5}{5} \qquad 7 = 6\frac{13}{13} \qquad 7 = 6\frac{10}{10} \qquad 6\frac{3}{3}$$

$$7 = 6 - ? \qquad 7 = 6 - ? \qquad 7 = 6 - ? \qquad 7 = 6 - ?$$

Rename the following whole numbers by filling the missing space

1) $7 = 6\frac{7}{}$	2) $7 = 6\frac{}{9}$	3) $7 = 6\frac{11}{}$	4) $7 = 6\frac{}{15}$
5) $9 = 8\frac{3}{}$?	6) $9 = 8\frac{3}{}$	7) $9 = 8\frac{3}{}$	8) $9 = 8\frac{3}{}$
9) $2 = 1\frac{7}{}$?	10) $2 = 1\frac{9}{}$?	11) $2 = 1\frac{}{6}$?	12) $2 = 1 - ?$

25) RENAMING FRACTIONS

1) To rename change the 5 into 4 +1	$5\frac{3}{7} = 4\frac{3}{7} + 1$
2) Change the 1 into 7/7 to match the denominator and add!	$4\frac{3}{7} + \frac{7}{7} = 4\frac{10}{7}$

Short Cut: More Examples:

Reduce the whole number by 1 and add the denominator to numerator to get the new numerator. Keep the denominator unchanged

2) $5\frac{3}{10} = 4\frac{10+3}{10} = 4\frac{13}{10}$ 2) $8\frac{2}{7} = 7\frac{9}{7}$ 3) $6\frac{3}{4} = 5\frac{7}{4}$

Rename the following mixed numbers or whole by reducing the whole numbers by 1. Use the short cut as in above examples!

1) $5\frac{4}{5}$ 2) $6\frac{1}{4}$ 3) $10\frac{2}{7}$ 4) 9

5) $8\frac{1}{13}$ 6) $4\frac{4}{5}$

7) 14   8) $8\frac{4}{9}$

9) $5\frac{6}{7}$ 10) $11\frac{9}{13}$ 11) $9\frac{2}{9}$ 12) $7\frac{7}{10}$

31

Simplify the fractions:

Example 1

Example 2

$$5 - \frac{3}{4} \downarrow \qquad 7 - 1\frac{3}{5} \downarrow$$

	Example 1	Example 2
1. Rename any whole number to match the denominator!	$5 = 4\frac{4}{4}$	$7 = 6\frac{5}{5}$
2) Subtract the fractions.	$4\frac{4}{4} - \frac{3}{4} = 4\frac{1}{4}$	$6\frac{5}{5} - 1\frac{3}{5} = 5\frac{2}{5}$

1) $12 - \frac{2}{5}$

2) $5 - \frac{2}{3}$

3) $5 - \frac{3}{10}$

4) $10 - \frac{8}{3}$

5) $4 - 1\frac{2}{3}$

6) $10 - 6\frac{1}{5}$

7) $11 - \frac{12}{15}$

8) $9 - \frac{5}{9}$

9) $12 - 4\frac{3}{10}$

10) $13 - \frac{6}{7}$

11) $2 - \frac{2}{3}$

12) $14 - \frac{3}{25}$

13) $17 - \frac{1}{5}$

14) $9 - \frac{5}{6}$

15) $8 - 3\frac{6}{11}$

27) SUBTRACT MIXED NUMBERS: THE RENAMING METHOD

1. **Step1.** See we couldn't subtract the numerators$(3-5)$	$5\dfrac{3}{6} - 1\dfrac{5}{6}$
2. **Therefore, rename (borrow) as in previous lesson**	$4\dfrac{3}{6} + \dfrac{6}{6} = 4\dfrac{9}{6}$
3. **Now, subtract and simplify**	$4\dfrac{9}{6} - 1\dfrac{5}{6} = 3\dfrac{4}{6} = 3\dfrac{2}{3}$

1) $5\dfrac{3}{5} - 3\dfrac{4}{5}$

2) $7\dfrac{1}{5} - 3\dfrac{2}{5}$

3) $9\dfrac{2}{9} - 5\dfrac{3}{9}$

4) $5\dfrac{1}{7} - 2\dfrac{3}{7}$

5) $14\dfrac{1}{13} - 1\dfrac{3}{13}$

6) $7\dfrac{3}{5} - 2\dfrac{4}{5}$

7) $14\dfrac{2}{8} - 1\dfrac{7}{8}$

8) $7\dfrac{5}{16} - 2\dfrac{7}{16}$

9) $5\dfrac{3}{20} - 1\dfrac{7}{20}$

10) $15\dfrac{3}{7} - 3\dfrac{4}{7}$

11) $9\dfrac{2}{5} - 3\dfrac{4}{5}$

12) $8\dfrac{2}{7} - 5\dfrac{3}{7}$

Step 1. See we couldn't subtract the numerators $(3-5)$	$7\dfrac{5}{8} - 3\dfrac{7}{8}$
Step2. Rename: add the denominator to numerator to get the new numerator. Keep the denominator unchanged	$7\dfrac{13}{8} - 3\dfrac{7}{8}$
Step 3: Now, subtract and simplify	$= 4\dfrac{6}{8} = 4\dfrac{3}{4}$

1) $7\dfrac{3}{5} - 3\dfrac{4}{5}$

2) $8\dfrac{1}{5} - 4\dfrac{3}{5}$

3) $9\dfrac{1}{8} - 5\dfrac{3}{8}$

4) $6\dfrac{3}{7} - 4\dfrac{5}{7}$

5) $9\dfrac{7}{10} - 1\dfrac{9}{10}$

6) $4\dfrac{3}{5} - 3\dfrac{4}{5}$

7) $12\dfrac{5}{8} - 5\dfrac{7}{8}$

8) $7\dfrac{5}{16} - 2\dfrac{7}{16}$

9) $4\dfrac{11}{20} - 1\dfrac{13}{20}$

10) $13\dfrac{3}{10} - 3\dfrac{7}{10}$

11) $9\dfrac{2}{7} - 3\dfrac{4}{7}$

12) $5\dfrac{3}{7} - 4\dfrac{5}{7}$

Simplify. Make sure to reduce and to rename the fractions as needed.

1) $\frac{1}{7} + 5$

2) $9 - \frac{4}{3}$

3) $\frac{11}{5} - \frac{3}{5}$

4) $7 - 5\frac{2}{7}$

5) $10 + \frac{11}{5}$

6) $\frac{7}{2} - \frac{3}{2}$

7) $\frac{12}{5} - \frac{6}{5}$

8) $6\frac{1}{8} - \frac{5}{8}$

9) $6\frac{2}{5} - 2\frac{4}{5}$

10) $8\frac{2}{5} + \frac{19}{5}$

11) $3\frac{4}{7} - 2\frac{1}{7}$

12) $\frac{5}{2} + \frac{12}{5}$

13) $10 - \frac{1}{3}$

14) $7\frac{11}{15} - 3\frac{13}{15}$

15) $11 - \frac{11}{3}$

30) ADD SIMPLE FRACTIONS: UNLIKE DENOMINATORS!

How to do it? Just create equivalent fractions with the same denominators! Make sure to simplify at the end!

Example 1:	Example 2:	Example 3:
$\dfrac{3}{10}+\dfrac{1}{5}$	$\dfrac{2}{3}+\dfrac{2}{9}$	$\dfrac{3}{14}+\dfrac{2}{7}$
$\dfrac{3}{10}+\dfrac{1^{\times2}}{5_{\times2}}$	$\dfrac{2^{\times3}}{3_{\times3}}+\dfrac{2}{9}$	$\dfrac{3}{14}+\dfrac{2^{\times2}}{7_{\times2}}$
$\dfrac{3}{10}+\dfrac{2}{10}=\dfrac{5}{10}=\dfrac{1}{2}$	$\dfrac{6}{9}+\dfrac{2}{9}=\dfrac{8}{9}$	$\dfrac{3}{14}+\dfrac{4}{14}=\dfrac{7}{14}=\dfrac{1}{2}$

1) $\dfrac{1}{4}+\dfrac{1}{2}$

2) $\dfrac{2}{3}+\dfrac{2}{9}$

3) $\dfrac{3}{10}+\dfrac{1}{5}$

4) $\dfrac{3}{4}+\dfrac{1}{8}$

5) $\dfrac{1}{14}+\dfrac{2}{7}$

6) $\dfrac{1}{5}+\dfrac{3}{15}$

7) $\dfrac{3}{10}+\dfrac{1}{2}$

8) $\dfrac{5}{8}+\dfrac{1}{4}$

9) $\dfrac{5}{14}+\dfrac{4}{7}$

10) $\dfrac{3}{5}+\dfrac{1}{20}$

11) $\dfrac{4}{5}+\dfrac{7}{50}$

12) $\dfrac{3}{7}+\dfrac{5}{21}$

13) $\dfrac{1}{4}+\dfrac{1}{12}$

14) $\dfrac{7}{5}+\dfrac{3}{25}$

15) $\dfrac{2}{5}+\dfrac{3}{20}$

31) ADD SIMPLE FRACTIONS: UNLIKE DENOMINATORS!

How to do it? Just create equivalent fractions with the same denominators!

Example 1:
$$\frac{3}{2} + \frac{1}{5}$$
$$\frac{3^{\times 5}}{2_{\times 5}} + \frac{1^{\times 2}}{5_{\times 2}}$$
$$\frac{15}{10} + \frac{2}{10} = \frac{17}{10} = 1\frac{7}{10}$$

Example 2:
$$\frac{2}{3} + \frac{1}{5}$$
$$\frac{2^{\times 5}}{3_{\times 5}} + \frac{1^{\times 3}}{5_{\times 3}}$$
$$\frac{10}{15} + \frac{3}{15} = \frac{13}{15}$$

Example 3:
$$\frac{3}{4} + \frac{2}{7}$$
$$\frac{3^{\times 7}}{4_{\times 7}} + \frac{2^{\times 4}}{7_{\times 4}}$$
$$\frac{21}{28} + \frac{8}{28} = \frac{29}{28} = 1\frac{1}{28}$$

1) $\dfrac{1}{4} + \dfrac{2}{5}$

2) $\dfrac{2}{3} + \dfrac{2}{5}$

3) $\dfrac{3}{5} + \dfrac{1}{4}$

4) $\dfrac{3}{4} + \dfrac{1}{5}$

5) $\dfrac{3}{7} + \dfrac{2}{5}$

6) $\dfrac{3}{8} + \dfrac{4}{5}$

7) $\dfrac{3}{4} + \dfrac{1}{5}$

8) $\dfrac{5}{6} + \dfrac{1}{4}$

9) $\dfrac{5}{9} + \dfrac{4}{5}$

10) $\dfrac{3}{4} + \dfrac{1}{7}$

11) $\dfrac{4}{5} + \dfrac{7}{6}$

12) $\dfrac{3}{8} + \dfrac{2}{3}$

13) $\dfrac{1}{8} + \dfrac{7}{5}$

14) $\dfrac{7}{10} + \dfrac{1}{3}$

15) $\dfrac{1}{10} + \dfrac{5}{6}$

It is helpful to learn how to find the **Least Common Multiple (LCM)** before we work on more complex fractions with unlike denominators.

How to find it? Just continuously multiply 1, 2, 3 etc. to each number until you get a match which is called Lowest common multiple (LCM). The first is done for you!

		List the multiples here				The LCM is
1)	6 and 8	6	12	18	**24**	**24**
		8	16	**24**	36	
2)	4 and 3					
3)	6 and 14					
4)	6 and 9					
5)	4 and 10					
6)	6 and 20					
7)	3 and 16					
8)	8 and 20					

33) FINDING THE LCM: DIVISION LADDER METHOD

How to do it? Just continuously divide by any common factor until you are done (until all bottom numbers turn into ones!). Then, multiply the numbers in the first column. See examples

Find the LCM of: 1) 6 and 8   2) 9 and 12

Example 1

	6	8
÷2	3	4
÷3	1	4
÷4		1

LCM= 2 × 3 × 4 = 24

Example 2

	9	12
÷3	3	4
÷3	1	4
÷4		1

LCM = 3 × 3 × 4 = 36

1. 8 and 4   2. 20 and 5   3. 40 and 8

4. 15 and 25   5. 10 and 15   6. 20 and 12

7. 7 and 28   8. 4 and 14   9. 30 and 10

10. 4 and 10   11. 6 and 8   12. 4 and 6

Simplify the fractions: $\dfrac{3}{4}+\dfrac{1}{6}$

1. **The LCM is 12.** So, make the denominators equal to 12 by multiplying appropriate numbers. ⟹	$\dfrac{3^{\times 3}}{4_{\times 3}}+\dfrac{1^{\times 2}}{6_{\times 2}}$
2) **Add & Simplify** as needed ⟹	$\dfrac{9}{12}+\dfrac{2}{12}=\dfrac{11}{12}$

***Make sure** to multiply the numerator the same number you multiply to the denominator*

1) $\dfrac{2}{3}+\dfrac{4}{7}$

2) $\dfrac{5}{8}+\dfrac{1}{2}$

3) $\dfrac{2}{7}+\dfrac{3}{5}$

4) $\dfrac{8}{9}+\dfrac{2}{3}$

5) $\dfrac{1}{6}+\dfrac{5}{8}$

6) $\dfrac{3}{4}+\dfrac{5}{9}$

7) $\dfrac{4}{5}+\dfrac{7}{10}$

8) $\dfrac{5}{9}+\dfrac{1}{6}$

9) $\dfrac{11}{12}+\dfrac{3}{8}$

10) $\dfrac{7}{8}+\dfrac{3}{5}$

11) $\dfrac{13}{15}+\dfrac{7}{10}$

12) $\dfrac{4}{9}+\dfrac{7}{12}$

13) $\dfrac{5}{12}+\dfrac{3}{8}$

14) $\dfrac{5}{8}+\dfrac{9}{14}$

15) $\dfrac{4}{14}+\dfrac{2}{21}$

1) $\dfrac{1}{4} + \dfrac{2}{5}$

2) $\dfrac{2}{10} + \dfrac{7}{15}$

3) $\dfrac{3}{10} + \dfrac{1}{20}$

4) $\dfrac{3}{4} + \dfrac{1}{10}$

5) $\dfrac{3}{8} + \dfrac{5}{6}$

6) $\dfrac{3}{10} + \dfrac{7}{15}$

7) $\dfrac{3}{5} + \dfrac{1}{4}$

8) $\dfrac{5}{8} + \dfrac{1}{12}$

9) $\dfrac{5}{14} + \dfrac{4}{21}$

10) $\dfrac{3}{5} + \dfrac{1}{7}$

11) $\dfrac{4}{7} + \dfrac{7}{6}$

12) $\dfrac{3}{4} + \dfrac{5}{12}$

13) $\dfrac{1}{4} + \dfrac{1}{12}$

14) $\dfrac{7}{3} + \dfrac{7}{5}$

15) $\dfrac{2}{15} + \dfrac{3}{20}$

16) $\dfrac{5}{6} + \dfrac{3}{10}$

17) $\dfrac{1}{8} + \dfrac{7}{10}$

18) $\dfrac{5}{12} + \dfrac{3}{16}$

Simplify the fractions: $\qquad 4\dfrac{1}{5} + \dfrac{1}{2}$

Step 1) Find the LCM	The LCM of 5 and 2 is 10
Step 2) Rename fractions with the common denominators or LCM.	$4\dfrac{1^{\times 2}}{5_{\times 2}} + \dfrac{1^{\times 5}}{2_{\times 5}}$
Step 3) Add the Fractions & Simplify.	$4\dfrac{2}{10} + \dfrac{5}{10} = 4\dfrac{7}{10}$

1) $5\dfrac{3}{5} + \dfrac{1}{2}$

2) $7\dfrac{1}{5} + \dfrac{2}{3}$

3) $6\dfrac{2}{3} + \dfrac{1}{9}$

4) $\dfrac{3}{4} + 2\dfrac{1}{5}$

5) $\dfrac{2}{5} + 1\dfrac{7}{8}$

6) $3\dfrac{3}{16} + \dfrac{1}{8}$

7) $5\dfrac{3}{20} + \dfrac{7}{10}$

8) $1\dfrac{3}{5} + \dfrac{4}{9}$

9) $\dfrac{1}{7} + 3\dfrac{1}{4}$

10) $\dfrac{1}{2} + 1\dfrac{2}{9}$

11) $3\dfrac{2}{5} + \dfrac{5}{30}$

12) $\dfrac{2}{5} + 4\dfrac{5}{6}$

Simplify the fractions:

$$4\frac{1}{5} + 5\frac{1}{2}$$

Step 1) Find the LCM	**The LCM of 5, 2 is 10**
Step 2) Rename fraction with the common denominators or LCM.	$4\frac{1^{\times 2}}{5_{\times 2}} + 5\frac{1^{\times 5}}{2_{\times 5}}$
Step 3) Add the whole numbers & Simplify.	$4\frac{2}{10} + 5\frac{5}{10} = 9\frac{7}{10}$

1) $5\frac{3}{4} + 3\frac{1}{2}$

2) $7\frac{1}{2} + 3\frac{2}{3}$

3) $6\frac{2}{5} + 4\frac{1}{9}$

4) $4\frac{3}{4} + 2\frac{1}{5}$

5) $5\frac{5}{6} + 1\frac{3}{4}$

6) $3\frac{3}{8} + 2\frac{1}{5}$

7) $5\frac{3}{4} + 1\frac{7}{10}$

8) $5\frac{1}{5} + 2\frac{4}{7}$

9) $7\frac{1}{6} + 3\frac{1}{4}$

10) $9\frac{1}{5} + 1\frac{2}{3}$

11) $8\frac{1}{10} + 1\frac{7}{30}$

12) $2\frac{2}{9} + 1\frac{5}{6}$

38) SUBTRACT FRACTIONS: UNLIKE DENOMINATORS!

Simplify the fractions: $\dfrac{3}{5} - \dfrac{1}{4}$

1. Find the LCM	The LCM of 5 and 4 is 20
2. Rename fractions with common denominator. 3) Make sure to simplify at the end.	$\dfrac{3^{\times 4}}{5_{\times 4}} - \dfrac{1^{\times 5}}{4_{\times 5}}$ 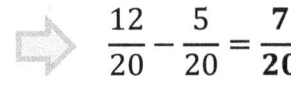 $\dfrac{12}{20} - \dfrac{5}{20} = \dfrac{7}{20}$

****Make sure** to multiply the numerator the same number you multiply to the denominator*

1) $\dfrac{2}{5} - \dfrac{2}{15}$

2) $\dfrac{5}{9} - \dfrac{2}{5}$

3) $\dfrac{3}{8} - \dfrac{1}{6}$

4) $\dfrac{3}{4} - \dfrac{1}{5}$

5) $\dfrac{5}{8} - \dfrac{1}{3}$

6) $\dfrac{5}{6} - \dfrac{3}{8}$

7) $\dfrac{6}{7} - \dfrac{1}{4}$

8) $\dfrac{5}{8} - \dfrac{1}{4}$

9) $\dfrac{5}{14} - \dfrac{2}{21}$

10) $\dfrac{9}{14} - \dfrac{5}{8}$

11) $\dfrac{11}{15} - \dfrac{5}{12}$

12) $\dfrac{3}{14} - \dfrac{3}{28}$

Simplify the fractions $5\frac{3}{5} - 4\frac{1}{2}$

Step 1) Find the LCM	Step 2) Rename fraction with the LCM.	Step 3) Subtract & Simplify
The LCM of 5, 2 is 10	$5\dfrac{3^{\times 2}}{5_{\times 2}} - 4\dfrac{1^{\times 5}}{2_{\times 5}}$	$5\dfrac{6}{10} - 4\dfrac{5}{10} = 1\dfrac{1}{10}$

**Remember to borrow or rename as needed: (You might want to review renaming lessons)

1) $5\frac{3}{5} - 3\frac{1}{2}$

2) $6\frac{1}{2} - 3\frac{2}{5}$

3) $9\frac{2}{3} - 5\frac{5}{9}$

4) $6\frac{3}{15} - \frac{1}{5}$

5) $7\frac{5}{6} - 1\frac{2}{3}$

6) $4\frac{2}{6} - 2\frac{1}{8}$

7) $4\frac{5}{12} - 1\frac{5}{8}$

8) $8\frac{3}{15} - 3\frac{4}{9}$

9) $6\frac{1}{4} - 1\frac{4}{7}$

10) $9\frac{1}{2} - 1\frac{2}{9}$

11) $3\frac{8}{45} - 1\frac{1}{9}$

12) $7\frac{2}{5} - 1\frac{5}{6}$

Example $8\dfrac{3}{5} - 1\dfrac{3}{4}$

1. Step 1: Make the denominators equal: use LCM	$8\dfrac{12}{20} - 1\dfrac{15}{20}$
2. Step 2: Since we can't subtract the 15 from the 12, we need to borrow (Rename) & simplify	$8\dfrac{32}{20} - 1\dfrac{15}{20} = 7\dfrac{17}{20}$

(You might want to review renaming lessons that we have worked earlier in the book)

1) $9\dfrac{3}{5} - 3\dfrac{2}{3}$

2) $6\dfrac{1}{12} - 5\dfrac{5}{6}$

3) $8\dfrac{2}{21} - 5\dfrac{4}{7}$

4) $6\dfrac{3}{12} - \dfrac{1}{4}$

5) $7\dfrac{5}{6} - 1\dfrac{5}{8}$

6) $5\dfrac{1}{16} - 2\dfrac{1}{4}$

7) $10\dfrac{3}{20} - 4\dfrac{2}{5}$

8) $9\dfrac{1}{5} - 3\dfrac{4}{9}$

9) $7\dfrac{1}{7} - 3\dfrac{1}{4}$

10) $8\dfrac{1}{4} - 3\dfrac{2}{3}$

11) $5\dfrac{2}{15} - 5\dfrac{1}{10}$

12) $9\dfrac{1}{6} - 1\dfrac{4}{9}$

Simplify fractions by performing the indicated operations

1) $3 + \dfrac{2}{3}$

2) $\dfrac{15}{8} + 5$

3) $\dfrac{5}{12} + \dfrac{1}{12}$

4) $\dfrac{5}{6} + \dfrac{7}{6}$

5) $19 - \dfrac{8}{13}$

6) $10 - \dfrac{5}{6}$

7) $\dfrac{8}{15} - \dfrac{3}{10}$

8) $7\dfrac{1}{7} + 3\dfrac{1}{4}$

9) $5\dfrac{5}{12} + 5\dfrac{3}{4}$

10) $8\dfrac{1}{5} - 3\dfrac{2}{3}$

11) $9\dfrac{1}{2} - 3\dfrac{3}{5}$

12) $7\dfrac{5}{12} - 2\dfrac{3}{4}$

13) $4\dfrac{1}{2} - 3\dfrac{2}{3}$

14) $9\dfrac{1}{5} + 3\dfrac{1}{4}$

15) $7\dfrac{5}{12} + 8\dfrac{1}{4}$

42) MULTIPLY FRACTIONS & WHOLE NUMBERS

Example: Simplify: $\frac{5}{12} \times 6$	**Step1:** Rename the whole number by placing a 1 under it and multiply $$\frac{5}{12} \times \frac{6}{1} = \frac{30}{12}$$	Step 2: **Simplify** $$\frac{30 \div 6}{12 \div 6} = \frac{5}{2} = 2\frac{1}{2}$$

It is quicker and easier **to use cross cancelation whenever you can!** The 6 simplifies with 12:

$$\frac{5}{\cancel{12}_2} \times \frac{\cancel{6}^1}{1} = \frac{5}{2} = 2\frac{1}{2}$$

Multiply & Simplify. Change any improper into mixed numbers.

1) $\frac{1}{2} \times 6$

2) $\frac{2}{5} \times 10$

3) $10 \times \frac{3}{4}$

4) $7 \times \frac{3}{14}$

5) $\frac{5}{36} \times 12$

6) $9 \times \frac{5}{12}$

7) $\frac{2}{7} \times 49$

8) $8 \times \frac{3}{16}$

9) $40 \times \frac{1}{8}$

10) $3 \times \frac{2}{15}$

11) $\frac{5}{80} \times 40$

12) $11 \times \frac{2}{33}$

Example: Simplify $1\frac{1}{12} \times 8$	Step1: Rename the mixed & the whole numbers & multiply: $\frac{13}{12} \times \frac{8}{1} = \frac{104}{12}$	Step 2: Simplify: $\frac{104 \div 4}{12 \div 4} = \frac{26}{3} = 8\frac{2}{3}$

Cross Cancelation method, **is quicker and easier to use:** The 8 and 12 each divides into 4:

$$\frac{13}{\cancel{12}_3} \times \frac{\cancel{8}^2}{1} = \frac{26}{3} = 8\frac{2}{3}$$

Multiply and rename any improper fractions into mixed numbers:

1) $3\frac{1}{2} \times 2$

2) $4 \times 2\frac{1}{6}$

3) $4 \times 1\frac{3}{4}$

4) $4\frac{1}{5} \times 5$

5) $1\frac{2}{5} \times 10$

6) $5\frac{2}{4} \times 2$

7) $3\frac{3}{16} \times 8$

8) $\frac{3}{20} \times 4$

9) $4\frac{1}{5} \times 3$

10) $3\frac{1}{5} \times 5$

11) $1\frac{1}{7} \times 21$

12) $2\frac{2}{9} \times 6$

Example: Simplify $\dfrac{3}{15} \times \dfrac{3}{5}$

| **Step1:** Multiply Numerators $3 \times 3 = 9$ | | **Step 2:** Multiply Denominators $15 \times 5 = 75$ | | **Step 3:** Divide the products and simplify. | $\dfrac{9 \div 3}{75 \div 3} = \dfrac{3}{25}$ |

Multiply fractions and simplify if possible:

1) $\dfrac{1}{5} \times \dfrac{5}{7}$

2) $\dfrac{5}{9} \times \dfrac{3}{7}$

3) $\dfrac{1}{3} \times \dfrac{3}{2}$

4) $\dfrac{4}{5} \times \dfrac{5}{8}$

5) $\dfrac{1}{12} \times \dfrac{3}{4}$

6) $\dfrac{7}{8} \times \dfrac{6}{11}$

7) $\dfrac{7}{2} \times \dfrac{3}{14}$

8) $\dfrac{3}{10} \times \dfrac{5}{6}$

9) $\dfrac{3}{4} \times \dfrac{4}{5}$

10) $\dfrac{5}{6} \times \dfrac{3}{10}$

11) $\dfrac{5}{14} \times \dfrac{7}{20}$

12) $\dfrac{7}{25} \times \dfrac{5}{2}$

45) MULTIPLY FRACTIONS: USE CANCELATION METHOD

Example : Simplify $\qquad$ 1) $\dfrac{5}{3} \times \dfrac{1}{10}$ $\qquad$ 2) $\dfrac{5}{3} \times \dfrac{1}{10}$

Reduce 5 with 10 and simplify:	Reduce 3 with 15 and simplify:
$\dfrac{\cancel{5}^{1}}{3} \times \dfrac{1}{\cancel{10}_{2}} = \dfrac{1}{6}$	$\dfrac{2}{\cancel{15}_{5}} \times \dfrac{\cancel{3}^{1}}{7} \times = \dfrac{2}{35}$

1) $\quad \dfrac{1}{15} \times \dfrac{5}{7}$

2) $\quad \dfrac{7}{12} \times \dfrac{8}{9}$

3) $\quad \dfrac{7}{15} \times \dfrac{5}{28}$

4) $\quad \dfrac{4}{9} \times \dfrac{3}{8}$

5) $\quad \dfrac{3}{22} \times \dfrac{11}{12}$

6) $\quad \dfrac{3}{27} \times \dfrac{9}{4}$

7) $\quad \dfrac{7}{5} \times \dfrac{3}{14}$

8) $\quad \dfrac{5}{9} \times \dfrac{3}{25}$

9) $\quad \dfrac{8}{5} \times \dfrac{1}{24}$

10) $\quad \dfrac{5}{9} \times \dfrac{8}{25}$

11) $\quad \dfrac{12}{13} \times \dfrac{5}{12}$

12) $\quad \dfrac{5}{14} \times \dfrac{7}{10}$

Example: Simplify: $\dfrac{7}{15} \times \dfrac{3}{35}$

Step1: Reduce 7 with 35	**Step2:** Reduce 3 with 15	**Step 3:** Multiply/Simplify
$\dfrac{\cancel{7}^{1}}{15} \times \dfrac{3}{\cancel{35}_{5}} \implies$	$\dfrac{1}{\cancel{15}_{5}} \times \dfrac{\cancel{3}^{1}}{5} \implies$	$\dfrac{1 \times 1}{5 \times 5} = \dfrac{1}{25}$

1) $\dfrac{2}{5} \times \dfrac{5}{6}$

2) $\dfrac{3}{5} \times \dfrac{5}{6}$

3) $\dfrac{4}{9} \times \dfrac{3}{8}$

4) $\dfrac{5}{6} \times \dfrac{4}{15}$

5) $\dfrac{7}{6} \times \dfrac{3}{14}$

6) $\dfrac{2}{3} \times \dfrac{9}{10}$

7) $\dfrac{5}{16} \times \dfrac{8}{25}$

8) $\dfrac{9}{35} \times \dfrac{7}{3}$

9) $\dfrac{7}{18} \times \dfrac{6}{14}$

10) $\dfrac{3}{8} \times \dfrac{8}{9}$

11) $\dfrac{8}{5} \times \dfrac{5}{8}$

12) $\dfrac{3}{77} \times \dfrac{11}{6}$

47) MULTIPLY MIXED NUMBERS WITH MIXED NUMBERS

Example: Simplify : $$5\frac{1}{4} \times 1\frac{1}{3}$$	**Step 1:** Change the mixed into improper fractions: $$\frac{21}{4} \times \frac{4}{3}$$	**Step 2:** Multiply and Simplify $$\frac{21 \times 4}{4 \times 3} = \frac{84}{12} = 7$$

✓ **It's easier to use cancelation:**

$$\frac{\cancel{21}^{7}}{\cancel{4}_{1}} \times \frac{\cancel{4}^{1}}{\cancel{3}_{1}} = \frac{7 \times 1}{1 \times 1} = 7$$

Use cancelation when possible to Simplify!

1) $3\frac{1}{2} \times 2\frac{4}{7}$

2) $2\frac{1}{2} \times 1\frac{3}{4}$

3) $4\frac{1}{5} \times 3\frac{4}{7}$

4) $5\frac{2}{4} \times 2\frac{1}{12}$

5) $3\frac{3}{16} \times 1\frac{1}{17}$

6) $4\frac{1}{5} \times 3\frac{1}{3}$

7) $3\frac{1}{5} \times 3\frac{3}{4}$

8) $2\frac{2}{9} \times 2\frac{7}{10}$

9) $1\frac{1}{5} \times 2\frac{1}{6}$

10) $1\frac{5}{9} \times 1\frac{2}{7}$

11) $1\frac{2}{5} \times 2\frac{1}{7}$

12) $7\frac{1}{5} \times 3\frac{2}{5}$

Simplify: make sure to change any improper fractions into mixed numbers

1) $\dfrac{12}{90} =$

2) $7\dfrac{1}{4} + 3\dfrac{1}{12} =$

3) $5\dfrac{1}{6} - 2\dfrac{1}{4} =$

4) $\dfrac{48}{9} \times \dfrac{3}{4} =$

5) $\dfrac{3}{8} \times 6 =$

6) $25 \times \dfrac{9}{10} =$

7) $3\dfrac{1}{6} \times 1\dfrac{1}{5} =$

8) $5\dfrac{1}{10} \times \dfrac{1}{3} =$

9) $\dfrac{2}{5} \times 6\dfrac{1}{4} =$

10) $2 \times 2\dfrac{2}{9} =$

11) $3\dfrac{1}{15} \times \dfrac{4}{5} =$

12) $2\dfrac{1}{6} \times 1\dfrac{5}{9}$

13) $2\dfrac{2}{5} \times 3\dfrac{1}{6}$

14) $1\dfrac{4}{5} \times 1\dfrac{3}{4}$

15) $2\dfrac{2}{5} \times 1\dfrac{6}{4}$

49) DIVIDE FRACTIONS BY FRACTIONS

Example: Simplify $\dfrac{2}{15} \div \dfrac{2}{5}$

1. **Keep** the first fraction as is. 2. **Change** the division sign into multiplication sign 3. **Flip** the divisor & simplify	$\dfrac{2}{15} \times \dfrac{5}{2} = \dfrac{10}{30} = \dfrac{1}{3}$

When it is possible, cross cancelation is easier

$$\dfrac{2^1}{15_3} \times \dfrac{5^1}{2_1} = \dfrac{1}{3}$$

Multiply fractions and simplify if possible:

1) $\dfrac{1}{5} \div \dfrac{5}{2}$

2) $\dfrac{4}{5} \div \dfrac{8}{5}$

3) $\dfrac{2}{7} \div \dfrac{3}{14}$

4) $\dfrac{5}{9} \div \dfrac{5}{18}$

5) $\dfrac{1}{2} \div \dfrac{5}{2}$

6) $\dfrac{3}{4} \div \dfrac{9}{4}$

7) $\dfrac{9}{8} \div \dfrac{9}{8}$

8) $\dfrac{7}{25} \div \dfrac{2}{5}$

9) $\dfrac{5}{9} \div \dfrac{3}{7}$

10) $\dfrac{9}{11} \div \dfrac{3}{22}$

11) $\dfrac{5}{10} \div \dfrac{1}{12}$

12) $\dfrac{5}{9} \div \dfrac{3}{7}$

Example: Simplify $\quad 1\frac{1}{15} \div 4$

1) Change the mixed number into improper fractions 2) Put one under the whole number 3) Change the division sign into multiplication 4) Flip & Simplify.	$\dfrac{16}{15} \div \dfrac{4}{1}$ $\dfrac{16}{15} \times \dfrac{1}{4} = \dfrac{16}{60} = \dfrac{4}{15}$	**When it is possible,** cross cancelation is easier $\dfrac{\cancel{16}^{4}}{15} \times \dfrac{1}{\cancel{4}^{1}} = \dfrac{4}{15}$

1) $3\frac{3}{8} \div 9$

2) $1\frac{1}{6} \div 7$

3) $\frac{5}{9} \div 5$

4) $\frac{3}{4} \div 5$

5) $1\frac{1}{4} \div 5$

6) $3\frac{2}{3} \div 22$

7) $4\frac{2}{3} \div 10$

8) $3\frac{1}{5} \div 4$

9) $\frac{7}{8} \div 49$

10) $8\frac{1}{6} \div 14$

11) $4\frac{1}{3} \div 26$

12) $2\frac{1}{4} \div 9$

Example: Simplify $4 \div 1\frac{1}{15}$

1) Put a one under the whole number 2) Change the mixed number into improper fraction	$\frac{4}{1} \div \frac{16}{15}$
3. Change the division sign into multiplication sign 4. Flip the fraction and simplify. Cross cancelation is quicker!	$\frac{4^{1}}{1} \times \frac{15}{16_{4}} = \frac{15}{4} = 3\frac{3}{4}$

1) $9 \div 3\frac{3}{8}$

2) $3 \div 6\frac{3}{4}$

3) $10 \div 4\frac{2}{3}$

4) $6 \div 8\frac{2}{5}$

5) $7 \div 1\frac{1}{6}$

6) $5 \div \frac{1}{4}$

7) $4 \div 5\frac{1}{5}$

8) $4 \div 9\frac{1}{3}$

9) $2 \div 3\frac{1}{4}$

10) $3 \div \frac{5}{4}$

11) $1 \div \frac{7}{8}$

12) $2 \div 7\frac{1}{4}$

52) DIVIDE MIXED NUMBER BY MIXED NUMBER

Example: Simplify $\qquad 1\dfrac{1}{15} \div 2\dfrac{2}{5}$

1. Change the mixed numbers into improper fractions.	$\dfrac{16}{15} \div \dfrac{12}{5}$
2. Change the division sign into multiplication sign 4. Flip the divisor (the second fraction) & simplify	$\dfrac{\overset{4}{\cancel{16}}}{\cancel{15}_3} \times \dfrac{\cancel{5}^1}{\cancel{12}^3} = \dfrac{4}{9}$

1) $5\dfrac{2}{3} \div 4\dfrac{6}{7}$

2) $1\dfrac{1}{5} \div 3\dfrac{1}{4}$

3) $1\dfrac{5}{9} \div 3\dfrac{1}{3}$

4) $6\dfrac{3}{4} \div 3\dfrac{3}{4}$

5) $1\dfrac{3}{4} \div 3\dfrac{1}{2}$

6) $1\dfrac{2}{3} \div 9\dfrac{4}{9}$

7) $4\dfrac{1}{5} \div 1\dfrac{1}{6}$

8) $2\dfrac{4}{15} \div 2\dfrac{7}{10}$

9) $1\dfrac{7}{18} \div 1\dfrac{3}{27}$

10) $9\dfrac{1}{8} \div 2\dfrac{5}{6}$

11) $9\dfrac{2}{3} \div 7\dfrac{1}{4}$

12) $1\dfrac{1}{4} \div 3\dfrac{1}{2}$

53) COMPLEX FRACTIONS

$Simplify: \dfrac{1}{5} \times 1\dfrac{1}{5} \times \dfrac{10}{3}$

1) **Change** any mixed into improper fractions	$\dfrac{1}{5} \times \dfrac{6}{5} \times \dfrac{10}{3} = \dfrac{60}{75}$	Cross cancellation is usually easier and quicker!
2) **Multiply (a)** all numerators & **(b)** all denominators	$\dfrac{60 \div 15}{75 \div 15} = \dfrac{4}{5}$	$\dfrac{1}{5} \times \dfrac{\cancel{6}^{2}}{\cancel{5}_{1}} \times \dfrac{\cancel{10}^{2}}{\cancel{3}_{1}} = \dfrac{4}{5}$
3) **Simplify**		

Note: *If there is division involved, be sure to flip. When both multiplication and division are involved work from left to right!*

1) $\dfrac{3}{4} \times \dfrac{1}{5} \times \dfrac{5}{3}$

2) $\dfrac{1}{4} \times \dfrac{2}{3} \times \dfrac{1}{2}$

3) $\dfrac{2}{3} \times \dfrac{3}{5} \times \dfrac{4}{9}$

4) $3\dfrac{3}{4} \times \dfrac{2}{5} \div \dfrac{1}{2}$

5) $\dfrac{10}{7} \times \dfrac{1}{3} \times \dfrac{21}{30}$

6) $2\dfrac{4}{5} \times 2\dfrac{1}{2} \times \dfrac{1}{3}$

7) $\dfrac{1}{5} \div 3\dfrac{1}{4} \times 1\dfrac{7}{8}$

8) $7\dfrac{1}{9} \div 2\dfrac{1}{4} \times \dfrac{1}{8}$

9) $5\dfrac{3}{5} \div 1\dfrac{3}{5} \times 5$

10) $3\dfrac{3}{5} \times 1\dfrac{1}{4} \div 3\dfrac{1}{2}$

11) $\dfrac{1}{3} \times 1\dfrac{4}{5} \times 5$

12) $7\dfrac{4}{5} \times 1\dfrac{6}{4} \div 3\dfrac{1}{2}$

1. **Read** the problem carefully and understand what is needed.

2. **Look for the signal words and decide what operations to use:** additions, subtractions, multiplications or division

Addition Signal words
Addition, add, plus, sum, more, increase, all, total, altogether….

Subtraction Signal Words
Subtraction, subtract, minus, difference, decreased, reduced, left over or remains, how many more…

Multiplication Signal Words
Multiply, Product, of, factor, how many times, twice…

Division Signal Words
Divide, average, shared evenly or equally, split equally, quotient…

Example 1. Two sisters wanted to add their shares in the new grocery. Aisha had $4\frac{1}{2}$ shares, Sahara had $2\frac{1}{2}$ shares. How many shares they altogether had?

    The signal word: altogether means addition:
$$4\frac{1}{2} + 2\frac{1}{2} = 6\frac{2}{2} = 6 + 1 = 7$$

Example 2: Ali used $5\frac{1}{4}$ cups of flour to prepare Sambusa. He had 9 cups of flour before. How much flour is left for him?

    The signal word: Left means Subtraction $\quad 9 - 5\frac{1}{4} = 8\frac{4}{4} - 5\frac{1}{4} = 3\frac{3}{4}$

Example 3: The height of the building was $12\frac{1}{4}$ feet high before it was doubled. How tall is the building now?

    The signal word: Doubled means multiply it by two!
$$12\frac{1}{4}\,ft \times 2 = \frac{49}{4} \times 2 = \frac{49}{2}ft = 24\frac{1}{2}ft$$

Work on these word problems. Get your hint from the signal words:

1) *Sareedo* took a math test and scored 88 out of 92 possible points. Write her score as a fraction and simplify.

2) Shakib works on math for **2 ¼** hours on Mondays, **$2\frac{5}{7}$** hours on Tuesdays and **$2\frac{7}{9}$** hours on Wednesdays. Which day he plays most?

        a) Monday   b) Tuesday   c) Wednesday

3) One hundred twenty five (125) students rode the bus today out of 350 students in the school.

    a) What fraction of students rode the bus today? Write it in simplest form

    b) What fraction of students didn't ride the bus? Write it in simplest form.

4) Jama eats 5/6 of cup of dates in the morning and 7/9 of cup in the evening. What is the total amount of dates that Jama used?

5) Before she started the gym, Samira was $180\frac{3}{4}$ pounds. She has lost $18\frac{1}{3}$ pounds since. How many pounds is he now?

6) *Lee* is cutting a curtain into three equal sizes. The curtain has a width of $5\frac{1}{4}$ feet. How long will each piece be?

7) Samsam has a board that is $13\frac{3}{4}$ ft long. If she cuts a piece that is $1\frac{1}{5}$ ft. and then another piece that is $2\frac{1}{3}$ ft., what is the length of the remaining piece?

8) Twenty seven students are in a classroom and each student had 5 books. The math teacher collected 1/3 of the books of the students. How many books the teacher has collected?

9) Anisa uses $\frac{3}{4}$ of a spoon of sugar for her breakfast cereals. This week she decreased it to 1/5 of a spoon. How much she has reduced her sugar? Write it in a simplified fraction.

10) Three students are making a class presentation. The first student took $4\frac{1}{5}$ minutes; the second took $7\frac{1}{3}$ minute and the last one took $5\frac{1}{4}$ minutes. Find the total time of the presentation.

11) Ayan had a 3 ¼ of the cake. She decided to give her sister ½ of what she has.

    a) How much she gave sister?

    b) What fraction of the cake is left for Ayan?

12) Farah has $15\frac{1}{4}$ yards of clothes to make curtains for his windows. He cuts $2\frac{5}{8}$ for his first window. How many yards is left?

13) The fox divided the hunt to the animals. She gave 2/5 of the meat to the lion, 4/7 to the hyena. How much meat the fox gave to both animals? Who got more?

14) Layla took a cab ride from the Airport to her home that was about $8\frac{3}{8}$ miles from the Airport. On her way She made a stop at a grocery store that was just $4\frac{1}{5}$ miles. How far is her home from the grocery?

15) Abooto was $10\frac{1}{2}$ pounds when she was born. Two years later she gained $30\frac{1}{4}$ pounds before she lost $\frac{1}{6}$ th of her total weight. What is her weight now?

16) Nawal's Cake recipe is shown on the right. Answer the questions:

a) How many ounces of eggs and flour has Nawal used?

b) How many ounces of flour, sugar and salt used?

Nawal's Cake Recipe	
Eggs	$\frac{3}{4}$ ounces
Flour	$10\frac{1}{3}$ ounces
sugar	$2\frac{1}{3}$ ounces
Salt	$\frac{1}{3}$ ounces

/

c) How many more ounces of flour were there than sugar?

1) $\dfrac{3}{90}$

2) $3\dfrac{5}{25}$

3) $\dfrac{104}{10}$

4) $9 - \dfrac{5}{9}$

5) $7 - \dfrac{5}{9}$

6) $\dfrac{5}{6} + \dfrac{1}{5}$

7) $7\dfrac{5}{6} - 1\dfrac{5}{8}$

8) $5\dfrac{1}{16} - 2\dfrac{1}{4}$

9) $1\dfrac{3}{5} + 3\dfrac{4}{9}$

10) $6\dfrac{1}{5} \times 5$

11) $6\dfrac{1}{5} \times 3\dfrac{1}{2}$

12) $3\dfrac{3}{5} \times \dfrac{1}{3}$

13) $\dfrac{2}{5} \times 6\dfrac{1}{4}$

14) $7 \div 3\dfrac{2}{5}$

15) $3\dfrac{1}{15} \div \dfrac{4}{5}$

16) $2\dfrac{5}{9} \div 12\dfrac{4}{9}$

17) $1\dfrac{3}{4} \div 2\dfrac{1}{2} \times 3\dfrac{1}{6}$

18) $2\dfrac{2}{5} \times 1\dfrac{4}{5} \div 3\dfrac{1}{3}$

56. ANSWERS TO SELECTED PROBLEMS

1) *Naming Fractions:*
   1) 3/8   3) 1/4   5) 2/11   10) 2/6

2) Identify
   1) I   3) I   5) I   7) M   9) P

3) Equivalent Fractions 1
   1) 16   3) 12   5) 8   7) 2   9) 5   11) 10   13) 32

4) Equivalent Fractions 2
   1) 5   3) 3   5) 19   7) 8   9) 6   11) 2

5) Compare Two Fractions: Method 1
   3) <   5) >   7) =   9) <   11) =

6) Simplify Proper Fractions: Method 1
   1) 1/2   3) 6/7   5) 2/5   7) 6/7   9) 1/4   11) 5/6   13) 5/7

7) Simplify Proper Fractions: (Method 2)
   1) 2/9   3) 1/5   5) 12/25   7) 4/11   9) 1/2   11) 7/6   13) 5/6

8) Simplify Mixed Numbers:
   1) $3\frac{1}{4}$ 3) $6\frac{2}{5}$ 5) $8\frac{4}{5}$ 7) $9\frac{1}{3}$ 9) $21\frac{1}{2}$ 11) $12\frac{1}{3}$ 13) $9\frac{2}{5}$

9) Change Improper into Mixed:
   1) $1\frac{1}{4}$ 3) $3\frac{3}{8}$ 5) $7\frac{1}{2}$ 7) $6\frac{1}{3}$ 9) $8\frac{7}{8}$ 11) $2\frac{9}{32}$ 13) $1\frac{8}{17}$

10) Simplify Improper Fractions:
   1) $2\frac{1}{2}$ 3) $1\frac{1}{2}$ 5) $2\frac{2}{3}$ 7) $5\frac{2}{3}$ 9) 9   11) $1\frac{3}{5}$ 13) $2\frac{1}{2}$

11) Simplify Improper Mixed Numbers
   1) $8\frac{1}{3}$ 3) 23   5) $11\frac{2}{3}$ 7) $8\frac{1}{2}$ 9) $14\frac{1}{5}$ 11) $3\frac{1}{3}$ 13) 9

12) Change Mixed Numbers into Improper Fractions:

1) $\frac{7}{2}$ 3) $\frac{33}{4}$ 5) $\frac{65}{7}$ 7) $\frac{31}{7}$ 9) $\frac{47}{14}$ 11) $\frac{23}{5}$ 13) $\frac{35}{8}$

13) Progress Check 1

1) $<$, 3) $>$ 5) $\frac{7}{40}$,   7) $\frac{4}{5}$ 9) $\frac{7}{2}$ 11) $\frac{65}{9}$ 13) $1\frac{1}{4}$ 15) $3\frac{3}{8}$ 17) $6\frac{2}{3}$ 19) $11\frac{1}{3}$

14) Add proper fraction and a whole number:

1) $2\frac{1}{4}$ 3) $3\frac{3}{10}$ 5) $3\frac{2}{3}$ 7) $6\frac{1}{5}$ 9) $10\frac{3}{11}$ 11) $9\frac{3}{11}$ 13) $6\frac{1}{12}$

15) *Add improper Fractions and a whole number:*

1) $3\frac{1}{3}$ 3) $9\frac{2}{3}$ 5) $4\frac{2}{3}$ 7) $8\frac{2}{5}$ 9) $13\frac{2}{3}$ 11) $10\frac{1}{6}$ 13) $9\frac{2}{5}$

16) Add proper Fractions with Like Denominators

1) $\frac{3}{5}$ 3) $\frac{5}{7}$ 5) 1   7) $1\frac{1}{2}$ 9) $1\frac{1}{7}$ 11) $1\frac{1}{2}$ 13) $\frac{1}{2}$

17) Add improper Fractions with Like Denominators

1) $2\frac{3}{5}$ 3) $4\frac{1}{5}$ 5) 4   7) 3   9) $3\frac{6}{7}$ 11) $1\frac{4}{5}$ 13) $2\frac{2}{5}$

18) Subtract Fractions with Like Denominators

1) $\frac{1}{5}$ 3) $\frac{3}{7}$ 5) $\frac{2}{9}$ 7) 0   9) $\frac{4}{7}$ 11) $\frac{1}{6}$ 13) $\frac{1}{5}$ 17) $\frac{5}{9}$

19) Progress Check:

1) $3\frac{3}{5}$ 3) $4\frac{3}{5}$ 5) $8\frac{2}{5}$ 7) $3\frac{3}{5}$ 9) $\frac{3}{7}$ 11) $8\frac{2}{3}$ 13) $13\frac{2}{3}$

20) Add Mixed Numbers: Like Denominators

1) $5\frac{4}{5}$ 3) $5\frac{3}{8}$ 5) $3\frac{9}{11}$ 7) $19\frac{1}{2}$ 9) $4\frac{3}{5}$ 11) $7\frac{1}{2}$ 13) $6\frac{14}{15}$ 15) $3\frac{4}{19}$

21) Add Mixed Numbers: Like Denominators II

1) 5   3) 18   5) $8\frac{2}{7}$ 7) $11\frac{2}{5}$ 9) 6   11) $7\frac{2}{13}$

22) Subtract Mixed Numbers: Like Denominators

1) $4\frac{1}{5}$ 3) $3\frac{1}{9}$ 5) $3\frac{1}{7}$ 7) $6\frac{1}{5}$ 9) $3\frac{3}{4}$ 11) 5

23) Subtract Mixed Numbers: Improper fraction Method

1) $1\frac{4}{5}$ 3) $5\frac{4}{5}$ 5) $3\frac{11}{13}$ 7) $2\frac{3}{8}$ 9) $3\frac{3}{5}$ 11) $3\frac{4}{5}$

24) Renaming Whole numbers and Fractions

1) $6\frac{7}{7}$　3) $6\frac{11}{11}$　5) $8\frac{3}{3}$　7) $8\frac{3}{3}$ etc　9) $8\frac{3}{3}$　11) $1\frac{6}{6}$

25) Borrow (Rename) Mixed Fractions

1) $5\frac{9}{5}$　　3) $9\frac{9}{7}$　5) $7\frac{14}{13}$　7) $13\frac{8}{8}$ etc　9) $4\frac{13}{7}$　11) $8\frac{11}{9}$

26) Subtract Whole number and Fractions;

1) $11\frac{3}{5}$　3) $4\frac{7}{10}$　5) $2\frac{1}{3}$　7) $10\frac{1}{5}$　9) $7\frac{7}{10}$　11) $1\frac{1}{3}$　13) $16\frac{11}{12}$　15) $4\frac{5}{11}$　17) $11\frac{2}{9}$

27) Subtract Mixed Numbers: Borrowing (Renaming) method

1) $1\frac{4}{5}$　3) $3\frac{8}{9}$　5) $12\frac{11}{13}$　7) $12\frac{3}{8}$　9) $3\frac{4}{5}$　11) $5\frac{3}{5}$

28) Subtract Mixed Numbers: Borrowing (Renaming) Short cut

1) $3\frac{4}{5}$　　3) $3\frac{3}{4}$　5) $7\frac{4}{5}$　7) $6\frac{3}{4}$ etc　9) 2　11) $5\frac{5}{7}$

29) PROGRESS :CHECK

1) $5\frac{1}{7}$　3) $\frac{8}{5}$　5) $12\frac{1}{5}$　7) $1\frac{1}{5}$　9) $3\frac{4}{5}$　11) $1\frac{3}{7}$　13) $9\frac{2}{3}$　15) $7\frac{1}{3}$

30) Add Simple Fractions Unlike denominators

1) $\frac{3}{4}$　3) $\frac{1}{2}$　5) $\frac{5}{14}$　7) $\frac{4}{5}$　9) $1\frac{13}{14}$　11) $\frac{47}{50}$

31) Add Simple Fractions Unlike denominators

1) $\frac{13}{20}$　3) $\frac{17}{20}$　5) $\frac{29}{35}$　7) $\frac{19}{20}$　9) $1\frac{16}{45}$　11) $1\frac{29}{30}$　13) $1\frac{21}{40}$　15) $\frac{14}{15}$

32) Find Least Common Multiple (LCM): Use Factor Lists

1) 24　2) 12　3) 42　4) 18　5) 20　6) 60　7) 48　8) 60

33) Find the Least Common Multiple (LCM): Use Division **Ladder**

1) 8　3) 40　5) 30　7) 28　9) 30　11) 60

34) Add Fractions: Unlike Denominators!

1) $1\frac{5}{21}$　3) $\frac{31}{35}$　5) $\frac{19}{24}$　7) $1\frac{1}{2}$　9) $1\frac{7}{24}$　11) $1\frac{17}{30}$

35) ADD FRACTIONS: MORE PRACTICE

1) $\frac{13}{20}$　3) $\frac{7}{20}$　5) $1\frac{5}{24}$　7) $\frac{17}{20}$　9) $\frac{23}{42}$　11) $1\frac{31}{42}$　13) $\frac{1}{3}$

37) ADD Mixed Numbers: Unlike Denominators!

1) $9\frac{1}{4}$ 3) $10\frac{23}{45}$ 5) $7\frac{7}{12}$ 7) $7\frac{9}{20}$ 9) $10\frac{5}{12}$ 11) $9\frac{1}{3}$

38) Subtract Fractions: Unlike Denominators!

1) $\frac{4}{15}$ 3) $\frac{5}{24}$ 5) $\frac{7}{24}$ 7) $\frac{17}{28}$ 9) $\frac{11}{42}$ 11) $\frac{19}{60}$

39) Subtract Mixed Numbers: Unlike Denominators!

1) $2\frac{1}{10}$ 3) $4\frac{1}{9}$ 5) $6\frac{1}{6}$ 7) $2\frac{19}{24}$ 9) $4\frac{19}{28}$ 11) $2\frac{1}{15}$

40) Subtract Mixed Numbers: Borrowing method

1) $5\frac{14}{15}$ 3) $2\frac{11}{21}$ 5) 6 7) $5\frac{3}{4}$ 9) $3\frac{25}{28}$ 11) $\frac{1}{30}$

41) Progress Check 4

1) $3\frac{2}{3}$ 3) $\frac{1}{2}$ 5) $18\frac{5}{13}$ 7) $\frac{7}{30}$ 9) $11\frac{1}{6}$ 11) $5\frac{9}{10}$

42) Multiply Fractions & a Whole Number: 1) 3 3) $7\frac{1}{2}$ 5) $1\frac{2}{3}$ 7) 14 9) 5 11) $2\frac{1}{2}$

43) Multiply Mixed Numbers & Whole Numbers

1) 7 3) 7 5) 14 7) $25\frac{1}{2}$ 9) $12\frac{3}{5}$ 11) 24

44) Multiply Fractions with Fractions

1) $\frac{1}{7}$ 3) $\frac{1}{2}$ 5) $\frac{1}{16}$ 7) $\frac{3}{4}$ 9) $\frac{3}{5}$ 11) $\frac{1}{8}$

45) Multiply Fractions: Use Cancelation

1) $\frac{1}{21}$ 3) $\frac{1}{12}$ 5) $\frac{1}{8}$ 7) $\frac{3}{10}$ 9) $\frac{1}{15}$ 11) $\frac{5}{13}$

46) Multiply Fractions: Use Double Cross Cancelation

1) $\frac{1}{3}$ 3) $\frac{1}{6}$ 5) $\frac{1}{4}$ 7) $\frac{1}{10}$ 9) $\frac{1}{6}$ 11) 1

47) Multiply Mixed Numbers with Mixed Numbers

1) 9 3) 15 5) $3\frac{3}{8}$ 7) 12 9) $2\frac{3}{5}$ 11) 3

48) Progress Check 5

1) $\frac{2}{15}$ 3) $2\frac{11}{12}$ 5) $2\frac{1}{4}$ 7) $3\frac{4}{5}$ 9) $2\frac{1}{2}$ 11) $2\frac{4}{15}$

49) Divide Fractions by Fractions

1) $\frac{2}{25}$ 3) $1\frac{1}{3}$ 5) $\frac{1}{5}$ 7) 1 9) $1\frac{8}{27}$ 11) 6

50) Divide Mixed or a Fraction by a whole Number

1) $\frac{3}{8}$ 3) $\frac{1}{9}$ 5) $\frac{1}{4}$ 7) $\frac{7}{15}$ 9) $\frac{1}{56}$ 11) $\frac{1}{6}$

51) Divide whole number by Mixed Number or a fraction

1) $2\frac{2}{3}$ 3) $2\frac{1}{7}$ 5) 6 7) $\frac{10}{13}$ 9) $\frac{8}{13}$ 11) $1\frac{1}{7}$

52) Divide Mixed Number by Mixed Number

1) $1\frac{1}{6}$ 3) $\frac{7}{15}$ 5) $\frac{1}{2}$ 7) $3\frac{3}{5}$ 9) $1\frac{1}{4}$ 11) 1

53) Complex Fractions

1) $\frac{1}{4}$ 3) $\frac{8}{45}$ 5) $\frac{1}{3}$ 7) $\frac{3}{26}$ 9) $3\frac{3}{5}$ 11) 3

54) Fraction Word Problems

1) $\frac{22}{23}$ 3) a) $\frac{5}{14}$ b) $\frac{9}{14}$ 5) $162\frac{5}{12}$ 7) $11\frac{11}{24}$ 9) $\frac{11}{20}$ 11) a) $1\frac{5}{8}$ b) $1\frac{5}{8}$ 13) $\frac{34}{35}$ 15) $33\frac{23}{24}$

55) Final fraction check: Simplify all

1) $\frac{1}{30}$ 3) $10\frac{2}{5}$ 5) $6\frac{4}{9}$ 7) $6\frac{5}{24}$ 9) $5\frac{2}{45}$ 11) $21\frac{7}{10}$ 13) $2\frac{1}{2}$ 15) $3\frac{5}{6}$

Beauty is not the clothes we wear,

But is knowledge combined with good character!